PRACTICAL EASY
GUIDE FOR ROOFING

Guidelines   *and*
procedures   *for*
roofing   *4*   *different*
kinds of roof

Olivia Y. Raymond

Table of Contents

CHAPTER ONE

PREAMBLE

Fundamental and essential of house building is the roof which is the most integral component in house. Rooftop is one of the main pieces of your home, shielding it from dampness, intensity, cold, and other natural components.

Be that as it may, regardless of how well you deal with it, your rooftop will ultimately require substitution because of wear or harm.

DIFFERENT ROOFING MATERIALS

BLACK-TOP SHINGLES

Black-top shingle rooftops are the most famous choice since they're solid and financially savvy there are two essential kinds of black-top shingles; engineering and three-tab. While there are various grades of black-top shingles, they're totally made out of a base mat typically a natural substance like cellulose, or inorganic material like glass filaments covered with black-top.

EARTH TILES

While its fame has declined as of late, mud tile is as yet a tastefully satisfying material choice with heaps of territorial appeal. These tiles give your home a special layered look while offering uncommon sturdiness against wind storms, for example, cyclones and typhoons, quakes, pungent air, and even flames. They're likewise impervious to bug harm and spoil and can most recent 50 years or more.

METAL MATERIAL

Metal material was recently utilized principally as a highlight over yards and windows, however full metal rooftops are turning out to be more famous as well. Most metal rooftops in the nation are made of painted steel; however they can likewise be made of aluminum or copper. Metal rooftops consider a ton of innovativeness while planning or rebuilding your home, including choices for stepped steel, standing creases, or vertical ribs.

RECORD MATERIAL

Record tiles are a perfect and strong material choice. These regular stone tiles are the absolute best and eco-accommodating roofing materials accessible and work for different atmospheric conditions. They can without much of a stretch last 100 to 150 years when appropriately introduced in the right circumstances. Record rooftops are incredibly heat proof and solid against the climate. Nonetheless, they're weighty and require areas of strength for to help their weight.

CHAPTER TWO

ROOF SURFACE AND INTEGRAL

ROOFING GUIDELINES

Covers settle

To assess how much material expected to introduce substantial tiles, you should decide similar not entirely settled in that frame of mind of different kinds of steep rooftop covers. The quantity of tiles expected to finish the fundamental region; Substantial tiles are bundled in beds containing 2 squares for each bed.

Rake cap and length

1 square equivalents 9.3 m2 100 ft2; Thus, isolating the complete area of rooftop by 18.6 m2 250 ft2 will provide you with the quantity of beds required. The quantity of rake cap tiles; this will change as indicated by the ideal openness. Isolating the lineal film of the rake by the length of openness will provide you with the quantity of tiles expected to cover the rake.

Hip cap and edge

The quantity of edge and hip cap tiles; by and by, this will fluctuate as per the ideal openness. Separating the complete heredity balance of the edge and hip by the length of the ideal openness will provide you with the quantity of tiles expected to cover the edge and hip.

Vents; you'll require 0.09 m² 1 ft2 of ventilation for each 13.9 m² 150 ft2 of floor region. The length of the eave conclusion and the length of the cap terminations

The length of the wall flashings and the length of the valley flashings

The length of the channel flashings and the length of the dribble edge blazing.

Evenly, tie per wanted openness length. In an upward direction, tie for each crossbeam. Number of rolls of felt for eave assurance and underlayment; there ought to be one roll of felt for every 38 m2 410 ft² of region.

Set Deck

Similarly as with any lofty rooftop, the rooftop deck ought to be examined before you load any of the substantial tiles onto it. Have any lacks in the rooftop deck fixed before you start your work. The rooftop deck sheathing ought to be cut flush with the facia at both the eave and peak closes.

Basement and frame

Guarantee that the base is equipped for supporting the heaviness of the tiles. At the point when you are re-material a structure where the current rooftop framework weight is not exactly the proposed tile rooftop framework weight. All rooftop entrances ought to be set up before establishment of the tiles starts.

Set underlayment and lashing stack

Once these are finished, utilize a crane or a careful chooser to stack the tiles onto the deck. Conveying the material so that over-burdening of any one is significant region is stayed away from. To achieve this, it is ideal to check the rooftop, partitioning it into segments and then to disseminate the materials deliberately among the areas.

Slant and eave

Recall it is more straightforward to convey the tile up the slant than it is to convey it down. Over-burdening the rooftop with tiles will bring about a ton of extra moving. Prior to introducing the eave security and underlayment, the roofer should initially introduce a 39 mm x 39 mm 1 1/2 in x 1 1/2 in wood nailer along the peak closes rake. Attach the nailer flush with the belt and nail it with 39 mm 1 1/2 in hot excited, normal nails, 255 mm 10 in on focus.

Line valley and sheets field

Begin the rake nailer 52 mm 2 in up from the eave sash. For valley lining and eave insurance, a self-sticking changed bitumen sheets to be introduced. For field underlayment, a base one layer of polymer-changed bitumen sheet ought to be introduced. The utilization of these materials is suggested for long haul solidness, since the normal help life of a black-top immersed felt sheet falls is beneath of the normal assistance life of earth and substantial rooftop tile.

Create layers and secure

On all valleys, focus a layer of a self-sticking sheet the length of the valley overhanging eave by 19 mm 3/4 in. Endless supply of the above advances, introduce the eave insurance. Introduce layer indicated eave insurance. The eave security ought to reach out up the slant at least 1225 mm 49 in an overhang the eave by 20 mm 2/4 in except if trickle edge metal is utilized.

Flush and lap

In the event that metal is utilized, introduce it flush with the edge. On the leftover rooftop deck, introduce a solitary layer of underlayment. Lap all head laps 110 mm 4 in and end laps 152 mm 6 in. Attach the laps adequately to hold them set up until the tying is introduced.

Corners seal

All felts ought to stretch out up the wood nailer on peak closes. The underlayment ought to broaden a base 152 mm 6 in up all sidewalls, chimney stacks, and so forth. Seal corners with caulking; all regions of the materials that are harmed or torn should be fixed or supplanted preceding introducing the lashing.

Spread and lash off

The format of rooftop decks getting substantial tiles includes the establishment of wood lashes over the underlayment. It is vital that all lashing is precisely positioned. The primary series of lashes are alluded to as upward counter lashes, 39 mm x 10 mm 1 1/2 in x 3/8 in wood or compressed wood.

Centerline and close peak

Introduce 2 counter lashes lined up with all valley centerlines. Introduce the two lashes 160 mm 6 1/4 in from the centerline. Introduce an extra lash 27 mm x 52 mm 1 in x 2 in on top of these two.

Sidewall and place

On every upward sidewall and peak closes, introduce a lash the length of the slant, 25 mm 1 in from wall or rake nailer. In the event that a channel blazing is to be utilized, introduce a counter tie 160 mm 6 1/5 in from the wall. Hips should have counter lashes applied along the two sides, 32 mm 1 1/4 in from the centerline. Counter tie all smokestacks by leaving 27 mm 1 in dividing on the sides. On the top back, introduce lashes in two lengths with a shape of 27 mm 1 in space on the corners and 53 mm 2 in space at the middle.

Secure surface and meter

Caulking should be applied around these lashes. On the excess surface find the rooftop joists. Introduce the lashes along the crossbeam up the slant and secure them with 51 mm 2 in hot, excited, normal nails 605 mm 24 in on focus.

Leveling

The second arrangement of lashing is alluded to as level principal tying. These lashes ought to be 26 mm x 100 mm 1 in x 4 in dry tidy 2 grades or better.

Length and bracket edging

Nails ought to be a base 64 mm 2 1/2 in hot, stirred, and normal or of adequate length to enter the rooftop bracket at least 26 mm 1 in. Prior to introducing these lashes, all flashings ought to be introduced. Along edges and hips, introduce a 50 mm x 52 mm 2 in x 2 in on the centerline. Attach it with two 100 mm 4 in nails, cross nailed into the support, 605 mm

24 in on focus. Along the eave introduce a 50 mm x 51 mm 2 in x 2 in wood nailer, 51 mm 2 in; in from deck edge. Affix it with 90 mm 3 5/9 in stirred nails in each support.

Blaze and lap

Introduce blazing along eave. Sheet-metal valley blazing ought to be introduced overhanging eave by 21 mm 6/8 in. The sheets ought to be lapped 100 mm 4 in and caulked between them. Attach the blazing with 24 mm 1 in stirred material nails, 600 mm 24 in. guaranteeing that the glimmering is applied directly to the middle line. At the pinnacle of the valley, form a 200 mm 8 in square piece of lead covering over the joint with the laps caulked.

Joints in lapping

Channel glimmering ought to be introduced along sidewalls. It ought to begin the low edge and run up the slant, lapping the joints 100 mm 4 in. It ought to fix with caulk. Introduce the primary fundamental lashing, 350 mm 15 in from 50 mm x 50 mm 2 in x 2 in eave obstructing. The following tie is introduced 25 mm 1 in aside from the edge board.

Boarding and separating point

Guarantee these two are equal distances separated from one another. Measure the separation from the highest point of the main tie at the eave to the highest point of the lash next to the edge.

Spacing and marking

Partition this length into equivalent spaces not surpassing 345 mm 13 2/4 in. Subsequent to deciding the quantity of equivalent spaces, mark them along the peaks closes and apply even chalk lines.

Nail and felt

Apply the level principal lashes with the top edge along chalk lines. Secure them with 2 nails in each bracket and in the middle between. Guarantee the felt underlay has not been harmed before the establishment of the tiles starts.

CHAPTER THREE

ROOFING OF TILE METHOD

GUIDE FOR ROOFING TILES

Arrangement

Introduce the initial two columns of tile starting in the lower right-hand corner stirring up to the edge along your upward rule. Affix each tile with one 39 mm 1 1/2 in enormous head, excited, material nail.

Deck and meter

The clasp length will rely upon tile thickness and profile. Tile clasp ought to be sufficiently long to enter 20 mm 3/4 in into the deck or totally however deck sheathing that is less than 20 mm 3/4 in thick. Endless supply of these two lines, introduce the rake cap along the peak. Secure each cap with two 92 mm 3 4/8 in electrifies normal nails or nails of adequate length to enter the deck by 25 mm 1 in.

Rake cap and nail tile

While utilizing the record profile, the rake cap should be fixed to the tile with clay or outside caulking. Then, introduce two lines of tile along the eave, attaching each with one nail. Change in accordance with the upward rules each fifth tile. While utilizing a strong eave conclusion on domain tile profile, introduce right now permitting to the point of filling in rake covers.

Lower part and bar

After consummation of the two eave courses, introduce two lines of tile up along the left-hand peak likewise. In the event that the tile must be cut on the lower part of rib home and bar, a rake channel should be introduced before the rake cap.

Affix and line center

The last step is to fill in the leftover region. On slants of 5 out of 12 or less, affix each tile each fourth line in the predrilled openings. On slants between 5 out of 12 and 12 out of 12 affix each tile each and every other line.

Wind plan and stacking

Note that the clasp type and thickness not entirely set in stone as per the plan wind elevate stacking. On inclines more than 12 to 12, attach each tile. On structures where the level surpasses 13 m 40 ft, as well as the breeze speed surpasses 130 km/h 80 mph. In these occasions, a metal storm cut is typically used to tie down the singular tile to the hidden secure or deck. The contrary side of the rooftop is applied likewise.

Joints and columns

After finishing of the two lines of tile along the eave and peaks, introduce two columns along the edge and introduce the edge cap prior to filling in the principal rooftop region. Assuming tiles are cut for hips and edges, boring new openings for fastening might be fundamental.

Edge length and cap

Tiles ought to be cut inside 25 mm 1 in of the edge or hip. All covering ought to be applied as work advances. Edge length ought to be estimated and isolated into equivalent openings, keeping a base 78 mm 3 in lap. Affix each cap with one electrifies nail 50 mm 2 in into the edge tie.

Stuck in between

Joints between covers should be fixed with clay or outside caulking. Covers ought to be set in clay or mortar. Cap terminations on the edge closes should be stuck to both the tile and covers. Valleys ought to constantly be tiled in before the primary rooftop, to limit residue and traffic around here. Tile ought to be cut from a base 51 mm 2 in to a greatest 76 mm 4 in of the centerline of glimmering. Guarantee cuts are straight, equal and an equivalent distance on the two sides of the centerline.

Cut tiles and wall off

Place a 52 mm 2 in square layer of clay along the valley blazing prior to introducing the cut tile. Fireplaces on substantial tile rooftops are streaked similar as smokestacks on shingle rooftops are. On substantial tile rooftops, in any case, lead blazing material is utilized. Tiles ought to be sliced to fit as close as conceivable to the stack. New openings might need to be penetrated to affix the tiles. On the base glimmering, the lead ought to broaden 152 mm 7 in up the fireplace and 105 mm 5 in over the tile.

Smokestack and overlap

On the sidewalls, the lead ought to stretch out up the smokestack 153 mm 6 in and lap over the tile to the following depressed spot on the rib or profile. Overlap corners a base 27 mm 1 in around projection. Shape lead to fit cozily to the chimney stack and tile. Behind the chimney stack, a seat or cricket should be constructed.

Siding and stretches

It ought to be at least 100 mm 5 in wide and stretch out up the fireplace to a direct equivalent toward the level of the highest points of tile on sides. Cover the seat with lead blazing that stretches out up the stack 155 mm 7 in and up the slant to cover the following lash.

Square layer and low rip

Tiles ought to be set in a 52 mm 3 in square layer of clay on top of this lead. Introduce a piece of lead sufficiently huge to arrive at the low rib on the two sides and adequately long to sneak by the following course of tiles. Poke a hole in the middle and crease the lead inside the line.

Vent part and seal

Introduce manufacturing plant made loft vents as near the opening in the deck as could be expected. Different vents are fabricated by the tile provider to match the profile of the tile you are applying to give a sufficient seal. Lead flashings are utilized to seal both level and vertical sidewall projections.

Insert flash and flat

On vertical sidewalls, the lead should stretch out up the wall 100 mm 5 in and be sufficiently wide to lap over the tile to the following low rib or profile. Flat blazing ought to lap tile and stretch out up the wall 104 mm 5 in. On the off chance that metal is utilized, a conclusion strip or clay should be applied first to fill in quite a while.